SAE EDGE™
RESEARCH REPORT

Unsettled Topics Concerning User Experience and Acceptance of Automated Vehicles

Amine Taleb-Bendiab, Ph.D.
Valeo North America, Inc.

EDGE DEVELOPMENT TEAM

Greg Brannon, *AAA*
Nicolas De Crémiers, *Navya*
Alex Epstein, *National Safety Council*
Rachel Forestier, *Valeo*
David Hofert, *Perrone Robotics*
Katharina Hottelart, *Valeo*
Kristin Kolodge, *J.D. Power*
Sheldon Russell, Ph.D., *Virginia Tech Transportation Institute*

SAE INTERNATIONAL®
Warrendale, Pennsylvania, USA

About the Publisher

SAE International® is a global association of more than 128,000 engineers and related technical experts in the aerospace, automotive, and commercial-vehicle industries. Our core competencies are lifelong learning and voluntary consensus standards development. Visit sae.org

SAE EDGE™ Research Report Disclaimer

SAE EDGE™ Research Reports focus on topics that are dynamic, in which knowledge is incomplete, and which have yet to be standardized. They represent the collective wisdom of a group of experts and serve as a practical guide to the reader in understanding an unsettled subject matter. They are not meant to provide a recommended practice or protocol. The experts have assembled as a community of practitioners to contribute and collectivize their thoughts and points of view; these are not the positions of the institutions or businesses with which they are affiliated nor one contributor's perspective advanced over other contributors. SAE EDGE™ Research Reports are the property of SAE International, and SAE alone is responsible for their content.

About This Publication

SAE EDGE™ Research Reports provide state-of-the-art and state-of-the-industry examinations of the most significant topics in mobility engineering. SAE EDGE™ contributors are experts from research, academia, and industry who have come together to explore and define the most critical advancements, challenges, and future direction in areas such as vehicle automation, unmanned aircraft, cybersecurity, advanced propulsion, advanced manufacturing, Internet of Things, and connectivity.

Related Resources

SAE MOBILUS® Automated & Connected Knowledge Hub
https://saemobilus.sae.org/automated-connected/

SAE Team

Frank Menchaca, Chief Growth Officer
Michael Thompson, Director, Standards, Information and Research Publications
Monica Nogueira, Director of Content Acquisition
Beth Ellen Dibeler, Product Manager
William Kucinski, Managing Technical Editor

Printed in USA

EPR2020012
ISSN 2640-3536
e-ISSN 2640-3544
ISBN 978-1-4686-0187-9

To purchase bulk quantities, please contact: SAE Customer Service

E-mail: CustomerService@sae.org
Phone: 877-606-7323 (*inside USA and Canada*)
+1-724-776-4970 (*outside USA*)
Fax: +1-724-776-0790

https://www.sae.org/publications/edge-research-reports

About the Editor

Amine Taleb-Bendiab, Ph.D. is the Research and Development Director for Valeo's Comfort and Driving Assistance (CDA) Business Group in North America. In this role, he leads the engineering organization in three focus areas of Valeo Intuitive Driving: Automated Driving and Parking, Connected Car, and Intuitive Controls - for the North American market. Author of several technical papers in leading industry and academic publications, Dr. Taleb-Bendiab earned his doctorate in Physical Chemistry from the University of Michigan, with focus in spectroscopy techniques.

Prior to joining Valeo's CDA Business Group, Dr. Taleb-Bendiab held several technical leadership positions at various automotive innovation suppliers in the area of active safety, advanced lighting, smart sensing devices, LED, and laser products. Prior to working in automotive, he worked in start-up and government research lab institutions. He builds on more than 25 years' experience in cutting-edge technologies.

Dr. Taleb-Bendiab is a 21-year member of SAE International and is Valeo's representative to the Automotive Information Sharing and Analysis Center (Auto-ISAC) advisory board. He has participated as an invited speaker and panelist on automated driving and user experience and has contributed as SAE technical paper reviewer around advanced driver assistance systems, human-machine interface, and smart sensing topics.

contents

metamorworks/Shutterstock.com

Unsettled Topics Concerning User Experience and Acceptance of Automated Vehicles

Abstract

Driver assistance and low-level automation are creating a pathway of experience and projected acceptance toward higher automation levels when there is a positive experience and consumers see value in the technology. Automated vehicles will provide new opportunities for how consumers spend their time as the driver role evolves with higher levels of automation. However, the basics must still be met - safety and reliability, along with comfort and convenience - in order to gain consumer trust.

This SAE EDGE™ Research Report aims at addressing the unsettled topic of user acceptance of automated driving and analyzing the user experience for a more intuitive and safe driving experience. First, we examine the requirements for safer driver/user engagement with driving for the various SAE automation levels. This is based on learnings from various recent qualitative and quantitative studies. Second, we analyze consumer sentiment toward automated driving - both consumer excitement about the perceived benefits and dislikes or concerns about the technology. In addition, we touch upon the findings from surveys about drivers' experience with advanced driving assistance technologies and how it applies to automated driving. Third, we highlight driver profiles observed during a user-centric experience in an immersive automated driving cockpit. As a result, we proposed - through a trust pyramid representation - a means of gradually increasing user trust through careful human-machine interface (HMI) delivery with appropriate levels of information that communicate safe driving. Ultimately, the goal is to build up user confidence in safe automated driving so that their time can be spent on entertainment or other non-driving tasks.

NOTE: SAE EDGE™ Research Reports are intended to identify and illuminate key issues in emerging, but still unsettled, technologies of interest to the mobility industry. The goal of SAE EDGE™ Research Reports is to stimulate discussion and work in the hope of promoting and speeding resolution of identified issues. SAE EDGE™ Research Reports are not intended to resolve the challenges they identify or close any topic to further scrutiny.

AMINE TALEB-BENDIAB, Ph.D.
Valeo North America, Inc.

Edge Development Team

Greg Brannon, *AAA*
Nicolas De Crémiers, *Navya*
Alex Epstein, *National Safety Council*
Rachel Forestier, *Valeo*
David Hofert, *Perrone Robotics*
Katharina Hottelart, *Valeo*
Kristin Kolodge, *J.D. Power*
Sheldon Russell, Ph.D.,
Virginia Tech Transportation Institute

ISSN 2640-3536

Introduction

State of the Industry

The idea of self-driving, or automated cars, might have been percolating since the invention of the automobile [1, 2]. Such an idea moved from science fiction to a more realistic feature at the start of the current millennium, thanks to the Defense Advanced Research Projects Agency (DARPA) Grand Challenge in 2004 and then the Urban Challenge in 2007 [3]. While these were proof-of-concept competitions, they proved to be an important catalyst and turning point for automated driving - moving the notion of self-driving cars from "someday" to "this is coming." Advanced Driving Assistance System (ADAS) sensors [4, 5] (e.g., camera, LiDAR, radar, and ultrasonic) - key sensing and perception elements for automated driving - became more readily available and benefited from the increased maturity from applications in other industries. The data generated by these various sensors are collected and processed by the onboard computing platform to plan and subsequently define vehicle-safe motion. With a processing footprint as small as a tablet computer, such electronic control units are currently increasing in performance into the tens to hundreds of trillions of operations per second [6]. The continued advances in processing power in the years to come are enabling the benefits of artificial intelligence in automated driving applications. Today, there are many companies worldwide developing automated cars. Players range from established carmakers to tech giants interested in disruptive technology to up-and-coming start-up mobility companies.

> *"As automated vehicle technologies permeate new vehicle offerings more broadly, consumers will begin to understand and enjoy the many benefits, including a reduction of accidents. But these technologies must deliver on their promise by responding appropriately to the many challenging situations vehicles and drivers face every day."*
>
> **Bruce Belzowski**
>
> ***Managing Director, Automotive Futures Group University of Michigan (Retired)***

The main goal of automated vehicles first and foremost is safety, ultimately leading to the eradication of crashes caused by human error. Other benefits can pertain to economics, environmental concerns, productivity, comfort, and accessibility by certain user groups who otherwise are not capable of driving a conventional car. Drivers commuting to work is a prime opportunity for the increased productivity and comfort. They could start their daily tasks, such as reading or writing emails, as soon as they get into the automated car heading to work. American drivers spent about 27 minutes driving each way to work in 2018, or on average about 230 hours commuting annually [7]. It is worth noting that U.S. consumers were recently polled by J.D. Power regarding the advantages of automated vehicles; the ability to perform other tasks while in transit was considered the third-highest ranked advantage to automated driving. The two other top advantages were transportation for those who are unable or unwilling to drive and less stress/road rage [8].

While the term automated vehicle might be loosely used for any vehicle with a driving automation function, SAE International released a revised SAE J3016™ standard [9] to help categorize and better define the different levels of automation. Six levels, L0 to L5, were defined. L0 corresponds to "no automation" - in other words, a manually driven vehicle - to L5, which is fully automated, or self-driving, vehicle. L1 is defined as "driver assistance" and from L2 to L4, automation goes from partial to high, accordingly. The level of note is L3, which corresponds to conditional automation where the driving responsibility can be exchanged between humans and automation systems. If the system encounters an unexpected situation during the automation phase and is not capable of handling it, driving responsibility would be transferred to the human driver. For this reason, the driver must remain vigilant in this level of automation. The six SAE levels, L0 to L5, are common language worldwide for any organization engaged with automated driving technology.

The Unsettled Questions about the Users of Automated Vehicles

As described above, automated driving promises to bring a much higher level of road safety and other benefits to the consumer and society. Let's look at the example of automated parking, which is a form of automated driving, and, depending on the parking technology system on board, can be ranked anywhere from L1 to L4 per SAE automation levels [9]. Such technology aims at bringing safety and convenience to the driver, resulting in lower stress levels for the driver or operator during the parking maneuver. In a study done by Reimer et al. [10, 11], it was demonstrated through objective physiological data and the self-reporting of test subjects that stress is reduced in the L1 semi-Automated Parking Assist (APA) versus a manual maneuver (Figure 1).

It is clear that automated driving aims at delivering a safe, stress-free, and enjoyable journey. In parallel with the technological development and optimization of the automated driving architecture, the question that our industry is also analyzing is the consumer acceptance of such technology. What user experiences should we provide to build consumer trust? The goal is to have the consumer focused on productive or pleasurable activities, such as performing work-related tasks, enjoying the view outside the vehicle, interacting with other passengers, or watching a movie. Consumer acceptance is strongly linked to the user's trust level.

The objective of this SAE EDGE™ Research Report is to analyze the state of consumer acceptance through recent surveys, followed by evaluation of user experience

FIGURE 1. **Heart rate before, during, and after parallel parking.**

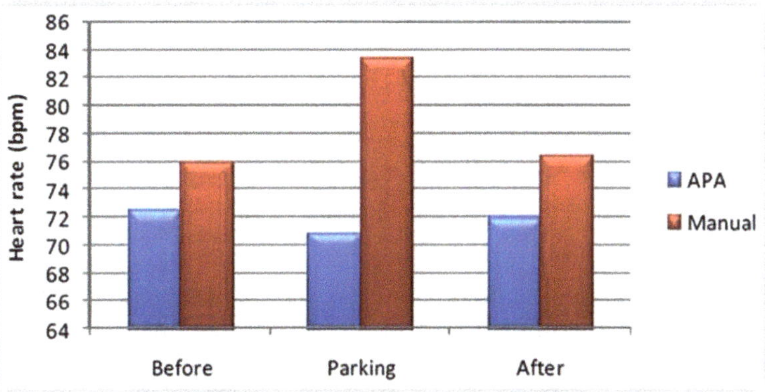

building blocks to enhance trust with increased levels of automation. But first, let's analyze the SAE automation level chart from the driver (or user) perspective along with safe driving expectations.

SAE Automation Levels: Driver Perspective

Driver Engagement Perspective

Looking closely at SAE J3016™ standard from the driver's perspective [9], there are three categories within the six levels of automation (Figure 2). From L0 through L2, the driver must be fully engaged in the driving task. More specifically, the Object and Event Detection and Response (OEDR) subtask of the Driving Domain Task (DDT) is the responsibility of the human driver for supervising the driving. That means "eyes-on-the-road," "hands-on-the-wheel," and "mind-on-the-driving-task." As we move to L3, the driver can be fully disengaged from the driving task while the vehicle is in the automated mode. Per SAE J3016™ definition, the L3 automated vehicle performs the entire DDT, including the OEDR subtask. However, the driver needs to be ready to take control if/when there is a need to switch back to manual driving. In this condition, the human driver is expected, in case of a DDT Fallback, to be receptive to the automated vehicle's issued request to intervene (RTI). At L4 and L5, the automated vehicle is fully responsible for the DDT as well as the DDT Fallback. A vehicle with automation levels L3, L4, or L5 is defined as one equipped with an Automated Driving System (ADS) [9]. Moreover, ADS in L4 or L5 is also defined as "driver-less operation" [9]; in which case the driver is completely disengaged from the driving function.

FIGURE 2. **SAE automation levels per SAE J3016™ versus driver engagement.**

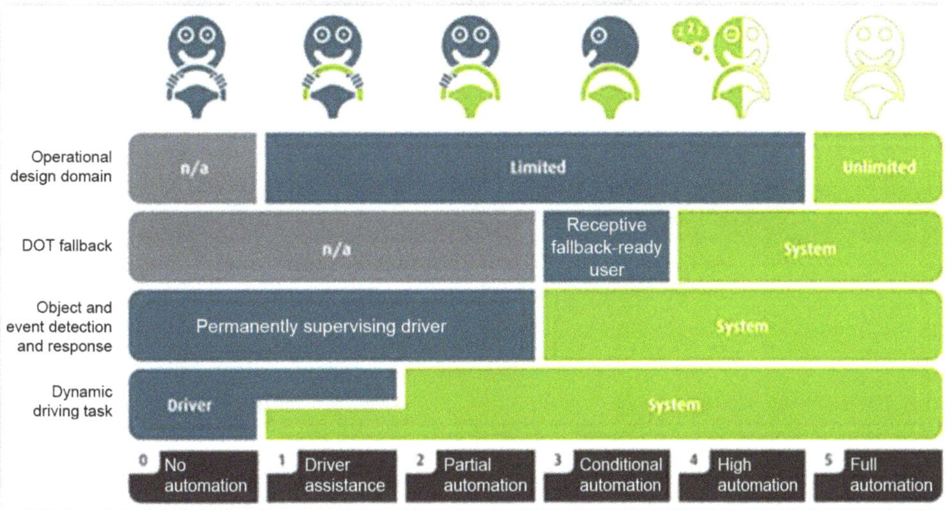

Safe Driving when Driver in Control: L0 through L2

Examining a 2018 report from the National Highway Traffic Safety Administration (NHTSA) on distracted driving and analyzing traffic crash data from 2016 [12], 9% of fatal crashes were reported as distraction-affected crashes; with 14% of these crashes, a cell phone was in use. Furthermore, a report from Virginia Tech Transportation Institute (VTTI) highlights that drivers using a handheld phone increased their crash risk by two to three-and-a-half times compared to "model" drivers (defined as alert, attentive, and sober) [13]. The report noted that any activity that places visual and manual demands on the driver, such as texting, browsing, or dialing a handheld phone, is the cause of a substantial increase of crash risk. On the other hand, talking on a voice-activated, hands-free device did not seem to have a detrimental effect on the cognitive state of the driver. In the author's own experience, even though the mind is distracted by talking with a hands-free device, the driver tends to keep their eyes on the road and keep their hands on the wheel. While the comparison between hands-free and handheld activities sought to emphasize a quantifiable detrimental impact on the driver's attention with a handheld phone, there are other causes of distractions, summarized in a study by VTTI, that would result in the driver not paying attention to the road [14].

VTTI has also performed naturalistic driving studies on either Adaptive Cruise Control (ACC) for longitudinal control or Lane-Keeping Assist (LKA) for lateral control, as well as L2 driving automation functions in recent years [15, 16]. While not in the true sense of SAE J3016™, for this report, we would label ACC or LKA as L1 driving automation. The goal was to shed light on driver behavior, performance, and habits while operating the vehicle in L1 and L2 states, using vehicles with driving automation systems available on the market at the time of the data collection. Focusing on the "eyes-on-the-road" and "hands-on-the-wheel" in this report, information on such activity has been extracted in these L2 Naturalistic Driving Studies (L2NDS). For example, looking at the results in Reference 15, thereafter refer to as NHTSA L2NDS, it is clear that during the L1 and L2 activated drives, the findings indicate the presence of non-driving tasks and are contrary to the SAE J3016™ expectations on the human driver at these automation levels. The presence of non-driving tasks was also present in L0 operation. Further, the user's hands on/ hands off the wheel was also monitored when the RTI in L2 operation was triggered by the vehicle's HMI. As shown in Table 1, 26.2% of the time the hands were not on the wheel. User hands being on the wheel was 73.8% during the RTI onset. In addition, the frequency of off-road glances was 13.2% in L2 mode, with 4.2% of these instances lasting longer than 2 seconds. These values do not necessarily suggest, within statistical significance, any increase or decrease of non-driving tasks going from L0 to L2 automation. However, these imply the importance of feedback for the "eyes-on-the-road" and "hands-on-the-wheel" in all automation levels.

Recommendations

1. Based on these findings in a manual drive (L0), it is important to guarantee "eyes-on-the-road" and "hands-on-the-wheel" at all times to ensure the driver is in control and, consequently, safer driving. Feedback through HMI or active sensing technologies - such as a driver monitoring system and steering wheel hand-grip sensing, among other sensing technologies - are very relevant.
2. The previous recommendation, "eyes-on-the-road" and "hands-on-the-wheel," would extend to L1 and L2, as the driver is still responsible for control of the vehicle and its safe driving. It should be noted that a joint focus group study by VTTI and Valeo in 2017 examined what happened when participants in the driver's seat were subjected to tasks in L2-equipped vehicles. The driver's tendency was to take their hands off the wheel and perform those tasks, leading them to look away from the road while the vehicle was in L2 operation, and thus apparently creating a sense of overtrust. The two vehicles used in this clinic were neither equipped with driver monitoring nor steering wheel sensing systems; thus no automated alert was given during the study. Therefore, the recommendation in item 1 must be applicable here as well for feedback to the driver, reinforcing the need to continue holding the wheel and maintaining a visual focus on the driving and the road ahead.

TABLE 1. Key driver and system interaction findings from the NHTSA L2NDS.

Category	Description	L0	L1	L2
Driver performance	Hands-**off**-wheel at time of RTI onset[1]	NA	NA	26.2%
Driver-system interaction	Non-driving task prevalence	66%	62%	60%
Eyes-off road glances - Non-driving task[2]	Glance instances (EORT)	11.7%	11.2%	13.2%
	Glances >2 seconds[3]	2.3 %	2.8%	4.2%

[1] Request to Intervene (RTI) is only available in L2 [15].

[2] The severity of "non-driving tasks" are judged through the Eyes-Off-Road Time (EORT) glances, which are known as surrogate for distraction.

[3] The 2-second threshold is based on NHTSA guidelines published in 2013 [17], which recommend a maximum time for a single glance at a device and away from the road.

TABLE 2. Automated to manual driving handover scenarios in L3 ADS.

Scenarios		Information	Warning[1]	Handover support	Driver in control	Intervention (MRM)[2]
Planned handover	Responsive driver[3]	45 seconds	5 seconds	10 seconds	Yes	No
	Unresponsive drive[4]	45 seconds	5 seconds	10 seconds	No	Yes
Unplanned handover[5]	Responsive driver[3]	-	5 seconds	10 seconds	Yes	No
	Unresponsive driver[4]	-	5 seconds	10 seconds	No	Yes

1 Reasonable audible and/or visual alerts per Thatcham Research recommendation [18]
2 Minimum Risk Maneuver
3 In this case, the driver is in charge and taking control.
4 In this case, the driver is not in charge and the system is increasing the alert level, visual, audible, haptic, and some other forms of stronger alerts per Thatcham Research recommendation [18].
5 For Unplanned Handover, there is no prior notice to the driver, i.e., Information Time = 0 second [18].

Safe Driving when Driver or System in Control: L3

Operating at L3 level is unique, such that scenarios where handover control from the ADS to the driver with or without advanced warning can occur. As a first priority, the ADS must be capable of performing a fallback Minimum Risk Maneuver (MRM) as highlighted in the Thatcham Research and ABI Report [18]. The term MRM is equivalent to Minimum Risk Condition defined in SAE J3016™ [9]. This is needed in the case where the driver is unable to take over control. In the Thatcham report, four potential scenarios are considered that are a combination of Planned/Unplanned Handover and Responsive/Unresponsive Driver as shown in Table 2. Figure 3 illustrates the chronological steps in the handover. Two studies, highlighted in the Thatcham report, have estimated the time it takes the driver to fully take back control [19, 20] anywhere from 12 seconds to 45 seconds. In a NHTSA report, the mean take-back control time was 2.3 seconds or 17 seconds, depending on whether the take-back control alert was an “Imminent” or “Staged” warning, respectively [21]. The Thatcham report suggests using a value of 15 seconds, where taking full control consists of the combined “Warning” plus “Handover Support” in Table 2.

Recommendations

1. To safely satisfy the requirements of L3, MRM capability must be available in an ADS.
2. It is also apparent that in L3, active feedback sensing is necessary to automatically confirm timely engagement and control by the driver. For example, feedback sensing should confirm “eyes-on-the-road,” “hands-on-the-wheel,” and “foot-on-the-gas-pedal.” Active eye-gaze monitoring, such as toward cluster/road/rearview mirrors as the driver gradually - though rapidly - resumes driving, could also be beneficial in the driver’s readiness to take control [20]. Although still in its infancy, cognitive sensing techniques that help define the driver’s state-of-mind could be beneficial as well.
3. Carefully designing the HMI according to intuitive concepts to ensure faster and seamless reengagement is an ideal course of action.

One example, a cockpit concept developed by Valeo named Mobius [22], was designed to address L2 and L3 automated driving. This fully digital and reconfigurable cockpit was equipped with a driver monitoring system and steering wheel sensing, and brought infotainment media into the driver’s cluster during automated driving. The steering wheel was equipped with active sensing to automatically detect if the hands were on or off the wheel. Also, the steering wheel was equipped with haptic touch screen switches to allow interaction with the infotainment media, as well as perform typing or phone-dialing tasks. The Valeo Mobius® is shown in Figures 4a and 4b, where Figure 4a highlights the cockpit when in manual driving mode with a typical cluster display, whereas Figure 4b highlights the

FIGURE 3. Automated to manual driving handover stages in L3 ADS [18].

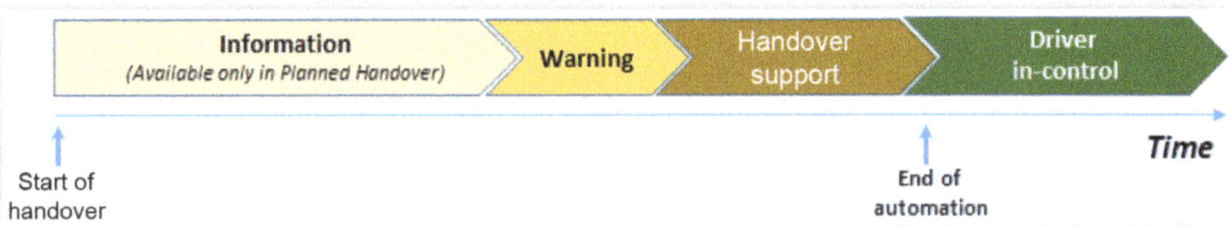

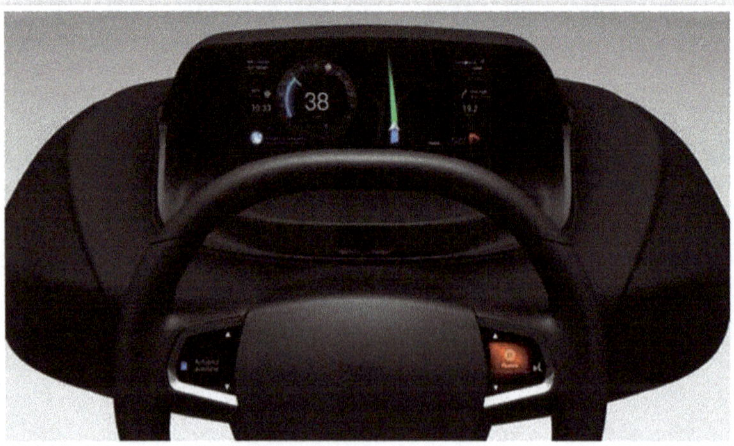

Reprinted with permission from Valeo

FIGURE 4A. Valeo Mobius® cockpit - manual driving view.

Reprinted with permission from Valeo

FIGURE 4B. Valeo Mobius® cockpit - automated driving view.

reconfigured cluster to facilitate the user experience interface in automated driving mode. This cockpit concept was designed with chimes and visual telltales to signal the driver about the need, for example, to take over control. Also, it was designed with the intent to keep the driver's eyes in the direction of the road when in manual mode and close to that viewing direction when in automated mode. Similarly, the steering wheel with hand-grip sensing and steering wheel sensing was designed to encourage the driver to have their hands on or in the vicinity of it when in manual and automated driving, respectively.

The Valeo Mobius® was evaluated and tested in a driving simulator at the Fraunhofer Institute for Industrial Engineering, utilizing 49 participants, performing the same tasks on a handheld smart device versus a mirrored one through the cockpit cluster [23] while the vehicle was in L3 automation. When an abrupt RTI was triggered requesting them to take control within 3.6 seconds, and if there was no reaction within the time window, a simulated emergency brake would stop the car. The results under the L2 automation condition with hands on the wheel and no infotainment on the cluster was used as a baseline. In this scenario, the mean takeover time was 1.85 seconds. This is consistent with the intervention time observed from the NHTSA L2NDS, in which the L2 mean takeover time with the hands initially not on the wheel was 1.79 seconds [15]. Although the Valeo Mobius® cockpit test was not fully representative of a dynamic driving environment, it still suggests a similar order of magnitude for the takeover time compared to the naturalistic driving results. When the Valeo Mobius® was in L3 mode, the mean takeover time was 2.95 seconds, since the driver was engaged in increased secondary tasks of a visual and cognitive nature. More importantly, using the Valeo Mobius® cockpit resulted in faster response time than using the handheld smartphone by 0.62 second, under L3 automation. To put things in perspective, 0.62 second corresponds to an extra 43 feet of stopping distance at 45 miles per hour.

In parallel to the reaction time study, the participants had the opportunity to provide their qualitative user experience feedback of the Valeo Mobius® cockpit compared to using a handheld smartphone while in L3 driving. The

FIGURE 5. Mcity Driverless Shuttle.

Reprinted with permission from Mcity

consensus was that the smartphone is still perceived as distracting even in automated driving. On the other hand, the cockpit provides a feel of both continued awareness of the traffic and safe driving [24].

Safe Driving when System in Control: L4 and L5

For the L4 and L5 ADS, the vehicle is expected to handle the driving without relying on the human driver (Figure 2). As the driving responsibility shifts to the system for driverless operation, the need for an intelligent monitoring system to guarantee "eyes-on-the-road" and "hands-on-the-wheel" are, in principle, not required anymore. Instead, one could imagine the monitoring function to rather focus on the well-being of the occupants and safety operator, when applicable. Other types of monitoring could be passenger count, seating position, type of passenger, safety monitoring of the occupant, etc. Today, research is at its infancy in defining the requirements for safe driving in L4/L5 automation from the point of view of occupant safety. However, it is worthwhile to highlight the Mcity case study on driverless operation shuttles. The Mcity Driverless Shuttle is the first driverless operation shuttle project in the USA, focused on user behavior and data collection. The L4 shuttle supplied by Navya, shown in Figure 5, operated at a maximum speed of 12 miles per hour. This experimental project aims at helping the industry understand the rider's experience of such technology as well as the interaction with other road users. The shuttle was equipped with interior cameras and audio recording for the analysis of parameters such as occupancy counts, emotional characterization, and seating position selection [25]. The goal of interior monitoring is to gain insight into user behavior and trust levels.

Acceptance of Automated Driving

Voice of the Consumers

There have been many surveys and clinics conducted in recent years designed to get a pulse on consumer sentiment regarding automated vehicles. They also seek to understand how easily consumers would embrace such innovation and have it as part of their commute modality. These studies have been conducted globally, exploring different demographic and generational views, urban and non-urban users, etc. With the goal to increase awareness and build trust and confidence in automated cars, SAE International launched a Demo Days event jointly with ADS-equipped vehicle partners. An example ADS vehicle from one of the event partners, Perrone Robotics, is shown in Figure 6. The SAE Demo Days event provided firsthand experience with the technology, offering more than 2,000 rides, and pre-ride and post-ride survey data were collected from nearly 1,400 participants between 2017 and 2019 around several cities [26]. This event was open to any consumer interested in taking part with the aim to increase awareness of this technology. Based on the post-ride survey, 82% of the participants were enthusiastic for ADS-equipped cars (Figure 7) - a 10% increase in the enthusiasm level from the pre-ride assessment.

The upbeat tendency from pre- to post-ride seems to be a common outcome from many different clinics, such as the driverless operation shuttle in the city of Las Vegas a few years back, where the rating by participants increased by 30% after the rides [27]. Knowledge is power. In a quarterly Mobility Confidence Index (MCI) Survey^{SM} conducted by J.D. Power, those who self-reported knowing a great deal about automated vehicles are more likely to purchase/lease

FIGURE 6. SAE Demo Days - partner ADS vehicle.

one compared to someone stating they don't know much at all about the technology [8] - 71% of respondents shared such intention, whereas this number dips to 26% when they lack knowledge of the technology. Test drives have the power to convince customers of the benefits of automated driving. A survey by Deloitte, where they polled German drivers, found that 68% of respondents are more likely to use automated vehicles if they were demonstrated under realistic conditions [28]. While there is clear consumer interest in the technology, there is still an underlying challenge of developing trust. For example, many consumers expressed a desire to take control at any time or lean toward a certain manufacturer brand with a superior safety track record.

"Control" Question Reviewing the SAE Demo Days survey [26], we can see in Figure 8 that 73% of participants would prefer to share control with the vehicle. This could imply a feeling from some of the users that driving remains a complex task and automated vehicles might not handle all driving scenarios as well as humans. Further, in the case of an emergency condition, 92% overwhelmingly favored the idea of having an "Emergency-Stop" feature in the ADS car. This feeling of being in control has also been echoed in the survey by Deloitte [28], where more than 90% wanted the ability to take control from the car at any time due to safety concerns; another 65%, of the Deloitte respondents are still worried about the technology not being reliable and that there could be a system malfunction. In the J.D. Power MCI Survey^{SM} [8], "Technology Failure or Errors" was the highest perceived disadvantage of automated vehicles.

The driving enjoyment and the user maintaining their driving skills could also suggest the desire of being in control of the driving. In fact, the J.D. Power MCI survey found that 32% of respondents dislike the idea of giving up the fun of driving [8].

Manufacturer Brand Relevance When posing the question about the trust level of automotive vehicles versus manufacturer label, the voice of consumers seems to differ between the reports from SAE [26] and Deloitte [28]. In the SAE results, while most participants were unsure, those that chose a manufacturer are split between "legacy manufacturer" and "new market entrant" (Figure 9a). What

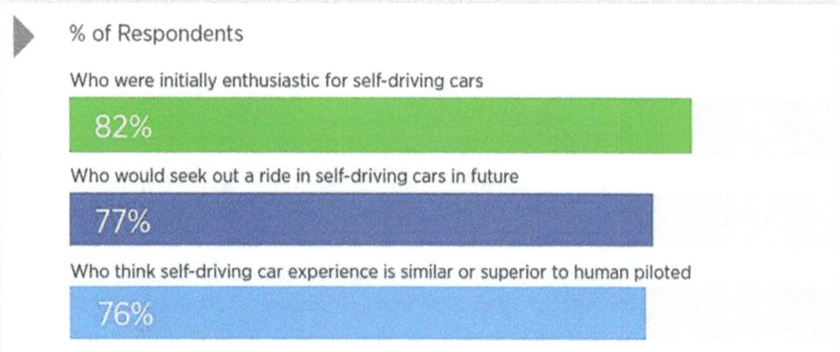

FIGURE 7. SAE Demo Days Survey - public enthusiasm regarding ADS.

FIGURE 8. SAE Demo Days Survey - sharing control with the car.

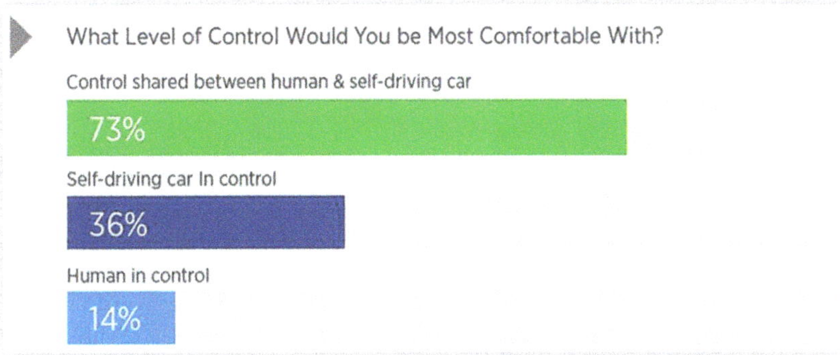

manufacturer to select was driven with almost equal weight based on the trust of the brand, innovation, and safety track record (Figure 9b). In the Deloitte results, consumers prefer the established premium brand manufacturers with 51% in favor. Leaning toward legacy premium brand signals the notion of a stronger track record for safety [28]. The SAE and Deloitte surveys were conducted in two different regions, the USA and Germany, respectively, and the cultural differences could have played a factor in the manufacturer brand preference.

Safety is a consumer expectation regarding automated vehicles. Being able to take control due to trust issues and favoring a technology from legacy and premium brand manufacturers seem to point to the perceived cautious consumer sentiments about ADS. Also, consumers are questioning the ability of ADS to independently handle all driving conditions and operate on the road with other road users [8, 26]. AAA has been tracking consumer sentiment regarding automated driving for the last four years [29]. While the fear sentiment to ride in an automated vehicle was at its lowest in December 2017 (63 percent), fear of riding in an automated vehicle increased by 10 percent in April 2018. Such fluctuation highlights the consumer cautiousness and potential influence from certain events surrounding the ADS vehicles [29].

Learning from Drivers' Experience with ADAS

Considering that common ADAS features are now available in as many as 80% of vehicles within several vehicle segments [30], driver experiences from using these features could be insightful toward user expectations of automated driving functions. It is apparent from several surveys, including some highlighted in this report, that acceptance of automated driving relies heavily on a positive test-driving/riding experience in an ADS or prior knowledge of the technology. For this reason, capturing driver sentiments about the ADAS features in their vehicles could provide meaningful feedback when defining both the advantages and the disadvantages of the technology [31]. Of course, perceived disadvantages would present an opportunity for our industry to improve and refine the technology to enable a superior and safer experience for the driver. Discussed below are some meaningful trends and sentiments that have surfaced from J.D. Power and AAA studies.

Annoyance from ADAS Alerts In the J.D. Power 2019 U.S. Tech Experience Index Study^{SM}, nearly 20,000 vehicle owners/lessees were asked about their user experience with various advanced technologies [32]. Among these technologies, they

FIGURE 9A. SAE Demo Days Survey - car brand preference for ADS.

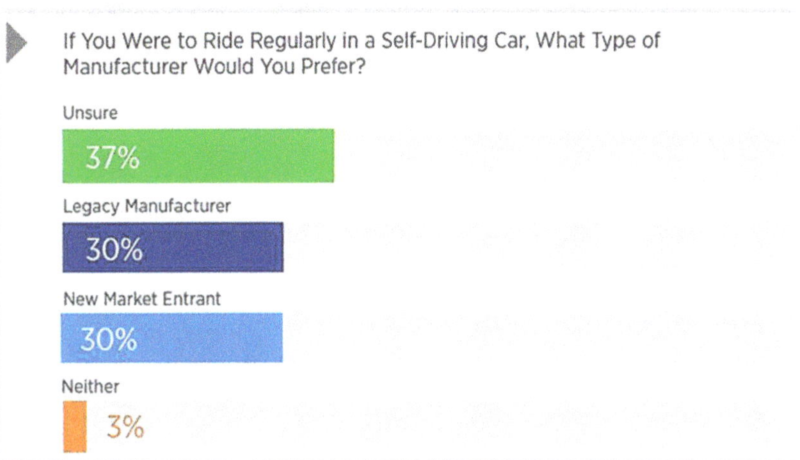

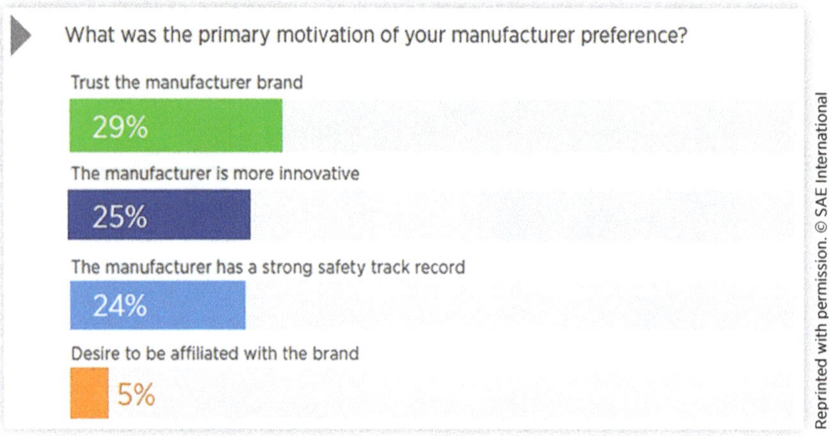

FIGURE 9B. **SAE Demo Days Survey - motivation for car brand selection.**

were asked about their experience with driver assistance systems. One concern this survey raised was that some of these ADAS features led to driver annoyance and frustration over the alerts, thus causing the driver to disable the feature. An example of such concern was the Lane Departure Warning (LDW) and Lane Keeping Assistance (LKA) systems, where, on average, 23% of respondents experienced this red-flag behavior (Figure 10). This behavior ranged from 8% to 30% depending on the vehicle make. Such variation is linked to performance differences, alert types, and overall alert strategy differences deployed across the industry. On average, 30% of consumers temporarily disabled their LDW/LKA system as a result. Moreover, only 63% of annoyed owners wanted this feature in their next purchase compared to 91% of those who were not annoyed. Along with the system's consistency perceived by the consumer, it seems clear that acceptance of new technologies such as these is complex and depends heavily on how HMI feedback messages are implemented.

Similar findings were observed in a study of 1,200 eligible respondents conducted by AAA and published a year earlier [31]. The LDW/LKA features were the most frequently deactivated, around 40% of the time, with annoyance as the main reason. AAA also examined five other driving assistance features and similar displeasure was expressed for the Front Collision Warning (FCW) and Automated Emergency Braking (AEB) systems (though with a lower rate of deactivation at 11%). The ADAS with the lowest deactivation rate were Blind Spot Warning (BSW) and Rear Cross Traffic Warning (RCTW). These were deactivated at a rate of 4% or less, suggesting a more intuitive nature of the feature and its benefits - although there was feedback that some deactivated the feature due to distractions in the case of BSW and false positives when towing a trailer or a boat in the RCTW case.

Misuse and Misunderstanding of ADAS Limitations Even if drivers are not annoyed by ADAS alerts, they still do not understand the limitations of such technology. AAA showcased such concerns in the previously mentioned study of drivers regarding ADAS technologies in newly purchased vehicles [31]. AAA found that about 80% of drivers did not understand the limits of their BSW function, such as how reliably it detects bicyclists and pedestrians and the limitations on detecting vehicles passing at very high speed. Furthermore, 25% of the drivers claimed they felt comfortable to rely solely on their BSW and RCTW systems. Similar results were seen on the limitations of FCW and AEB. Perhaps more concerning was the lack of understanding of the difference between these two features.

Confusion of ADAS Technology Naming Another factor leading to confusion of consumers and potentially false expectation among drivers are the various differentiating

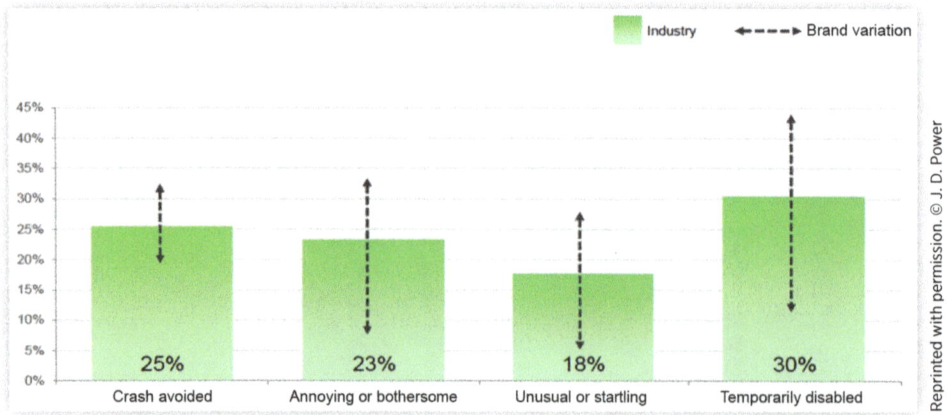

FIGURE 10. **ADAS trust level in LDW/LKA systems - J.D. Power 2019 U.S. Tech Experience Index Study℠.**

TABLE 3. Number of unique marketed names versus ADAS feature.

ADAS Feature	Unique Names
Adaptive Cruise Control	20
Lane Keeping Assistance	19
Blind Spot Warning	19
Rear Cross Traffic Assistance	14
Automatic High Beams	18
Semi-Automated Parking Assist	12

brand names for the ADAS features as pointed out in testimony by the National Safety Council (NSC) officials [33, 34]. It is unsettling that the same ADAS feature has between 12 to 20 different brand names as discovered in AAA's study [30] (Table 3). This data is based on 34 vehicle brands sold in the USA. This may not seem very important, but the variety of names has caused confusion about the intended ADAS function and is potentially misleading to the consumer.

For this reason, NSC called for a standardized nomenclature [34] and teamed up with AAA, Consumer Reports, and J.D. Power [35] in recommending common naming for advanced driving assistance technologies. Further, this initiative was endorsed by the U.S. Department of Transportation [36]. The goal of standardized naming is to bring more clarity to the consumer and - hopefully - gain industry and market acceptance. Logically, this concept might also be helpful to a consumer of automated driving vehicles. SAE International defined in J3016™ [9] the different automation levels and expectations for each level (Figure 2) - the goal is to avoid the misuse of the terms "autopilot" or "autonomous car." However, in the past year or so, unofficial terms "L2+" and maybe "L2++" started to surface and could potentially cause another naming confusion since these terms are not officially defined in SAE J3016™.

While not an exhaustive analysis about the undesirable trends impacting a positive driver experience using ADAS, the above categories (i.e., ADAS feedback annoyance, its limitations, or various naming for the same feature) are clear examples of areas that need to be refined in order to build a stronger foundation in the consumer acceptance of ADS as well.

User-Centric Analysis toward the Trust of Automated Driving

In a continued effort to understand the driver's expectations for automated driving, Valeo conducted a user-centric study analyzing driver experience when in a cockpit with simulated L3-like features (i.e., Traffic Jam Assist, Highway Pilot Assist, Automated Parking, etc.). The cockpit was equipped with technologies and features similar to those described for the Valeo Mobius® to enable switching from manual to automated driving mode, and vice versa. In addition, the cockpit was equipped with a Heads-Up Display (HUD), multi-display dashboard, and camera monitoring system. The cockpit was reconfigurable and allowed the driver to access entertainment media while the simulated vehicle was in automated driving mode.

The study was conducted between 2015 and 2017 in the USA, Germany, France, and Japan. Nearly 100 participants were assessed, with a mix of gender and age groups ranging from the mid-twenties to fifties. The selection criteria were focused on drivers - many of which owned cars with ACC and/or HUD system. As noted previously, familiarity with ADAS was potentially an enabler for awareness and trust in automated driving; thus the rationale behind selecting participants who owned a vehicle fitted with ACC. The selection of those who owned cars with HUD was intended to provide insight on drivers' awareness and reaction to newer user experience technologies. The clinic consisted of three phases:

Phase 1: Pre-clinic Interview - Collect insights on their driving habits, familiarity and experience with ADAS, and awareness as well as attitude toward HMI and automated driving.

Phase 2: Cockpit Experience - The test phase, behind the "wheel," and direct feedback. A picture of the cockpit is shown in Figure 11. Note: An early version of this cockpit was also used at the beginning of this research project.

FIGURE 11. Valeo Mobius® immersive cockpit.

Phase 3: Post-clinic Interview - Sharing session by participants of their experiences and expectations, including HMI optimization with respect to driving situations they faced.

One of the outcomes of this clinic was the discovery of four main driver groups. Each group had specific expectations from their vehicle, view on ADAS, and familiarity as well as proficiency with new technologies either in the car or in a consumer product. The outcomes are further described in the following sections.

Driver Profile Classification

Based on participant's feedback from the pre-clinic interview, four noticeable driver profiles were identified: Anxious, Cautious, Confident, and Sporty (Table 4). For example, those in the Anxious and Cautious profiles placed the utmost importance on safety. Their driving typically involved daily trips around town to run errands, commute to work, and driving kids to school or other activities; their technology knowledge was limited. On the other hand, those classified as Confident are very experienced drivers seeking comfort in their car. They enjoy driving in urban or long-distance trips and are very in tune with technology. However, they are not necessarily early adopters. Whereas the Sporty participants possessed very high confidence and skills as drivers and looked for a thrill from driving, while also enjoying new technologies in their cars. Interestingly, these same four driver profiles, or a subset of them, were observed in all study countries.

If we just focus on the technology savviness and the desire to drive, these profiles are classified as seen in Figure 12. The trend toward technology fan in the vertical axis would suggest self-confidence and trust in using newer technologies and, in the horizontal axis, when trending toward the pleasure of driving, would point to self-confidence behind the wheel and enjoyment of driving. With the Anxious and Cautious drivers putting safety as their primary focus, the Confident drivers are looking for comfort while driving. Finally, the Sporty drivers are looking for the pleasure of driving. Consequently, these various driver profiles' expectations in terms of ADAS usage are varied. Their respective focus of "safety above all," "comfort above all," or "pleasure above all" is defining how they would like to interact with the car when using ADAS features.

For instance, Anxious drivers seek to feel safe in their driving; however, too much information presented to them in the cockpit, such as alerts from a given ADAS feature, could be considered startling or even distracting. A Cautious driver wants intuitive ADAS features as they also expect safe driving, and the information presented to them through the cockpit must be easy to understand. For the Confident driver, while not early adopters of technology, they are very familiar with it and will adopt it if convinced that it will bring them safety and comfort benefits. Finally, Sporty drivers are quite knowledgeable and enjoy new technologies including ADAS, though mostly from an emotional benefit since they believe that it is their responsibility for driving and being in control. Each driver profile view on ADAS is summarized in Table 5.

HMI Expectations

Pre-clinic interviews and cockpit hands-on experience while the car drove in various simulated traffic and maneuver scenarios provided data to decipher each identified profile expectations for HMI (Table 5). Some of the findings are not that surprising. For example, the Anxious driver expected very little distraction while driving. Too much information from the dashboard becomes overwhelming. Whereas the Cautious driver is open to HMI feedback that complements their own senses. The Sporty and Confident drivers are more knowledgeable on the technologies, so their demand for HMI relates to their driving habits. On one hand, the Sporty driver wants that

TABLE 4. The four user-centric driver profile characteristics.

Anxious	Cautious
• Nervous driver, very guarded	• Vigilant driver, defensive driving
• Need-base relationship to the car	• Utilitarian relationship to the car
• Looking for safety, prevention, reassurance	• Looking for risk mitigation
• Daily local driving: work or errands	• Daily local driving: work or errands
• Not at ease with technology in the car	• Interested in technology in the car
• Mostly female and older	• More female and broad age spectrum
Confident	**Sporty**
• Experienced driver, self-confident	• Very confident driver, skilled
• Focus on stress-free and relaxed driving	• Strong bond with their car
• Car perceived as extension of home comfort	• Looking for the thrill from driving
• All types of driving with many long trips	• Daily commute, pleasure trips, and long trips
• Pragmatic use of new technology	• Tech geek and early adopters of technologies
• Mostly male and older	• Gender balanced with broad age spectrum

FIGURE 12. **Driver profiles in relation to driving and technology.**

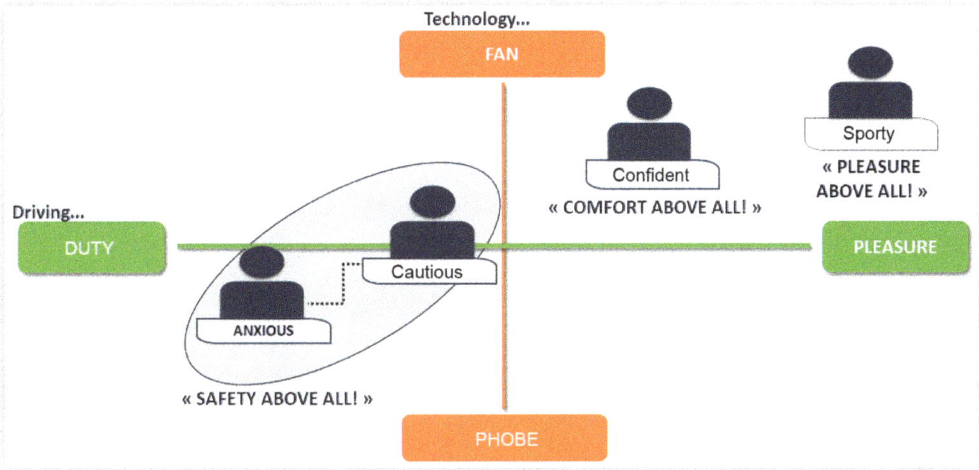

state-of-the-art technology to deliver on-demand, individualized features. On the other hand, the Confident driver expects their HMI to deliver a clear assessment and anticipation of any danger ahead or when delegating control to the ADS.

Attitude toward Automated Driving

One principal outcome of this study is to validate the link between experience with ADAS and the ability to trust an automated car. As we recall, many of the participants owned cars equipped with ACC. Further, drivers' hands-on experience with different simulated driving scenarios from L0 to L3, including transitions between automated and manual modes, helped anticipate their sentiments toward automated driving. As shown in Table 5, both the Anxious and Cautious drivers are rather reluctant to see themselves in an automated vehicle. Their concern is whether it is safe. They have less understanding of the advanced technology, leading to fears of malfunctions or accidents. The Sporty driver wishes to perform the driving themselves. They might

TABLE 5. **User-centric findings for each driver profile.**

Automotive Technology Knowledge	Driver Profile	View on ADAS	Expectation for HMI		View on Automated Driving
High	**Confident**	**Positive attitude** Experienced with the technology and a must-have if perceived beneficial	**Next Tech. Curiosity** Car understands and anticipates driver needs	&	*This driver shows openness to ADS, perceives the benefits, and can project themselves in ADS, especially the ones who used ACC*
	Sporty	**Mixed opinion** Limited use as "driver should be in control", and as tech-savvy emotional benefit prevails for ADAS	**Customization** Highly skilled driver looking for more Individual settings of ADAS		*This driver prefers to stay in control as it is fun to drive*
	Cautious	**Basic knowledge** Strong interest in taking advantage of ADAS	**Auditory feedback** Smart reminders to supplement own senses		*Both driver profiles are less capable to project themselves in ADS. Awareness is rising but fears supersede the benefits*
Low	**Anxious**	**Limited knowledge** Low interest and confidence in ADAS	**Less is better** Stay focused on driving. Override should be easy and prompt		

acquire the technology as a fun thing to have - but not necessarily to use, since they believe they should be in control. They could still be open to automation in specific driving situations such as being stuck in traffic or letting the car park itself. The Confident drivers showed openness to automated driving. They grasped its benefits such as safety, efficiency, and comfort, but still needed to be reassured with facts and data on the system. Some drivers' high-confidence level is based on their association with certain car brands that they perceive as high quality; as a result, they can project themselves more easily in an automated vehicle. Moreover, the experienced drivers who drove vehicles with ACC were the most enthusiastic about ADS.

Driver Trust in Automated Driving

This study attempted to understand if there were any perceived fears voiced by the participants. It also sought to capture through participant interviews the desired activities that could be performed during the automated drive.

The Drawback Perception There were two primary drawbacks voiced about automated cars. First, low confidence in the reliability of the technology - and second, the fear of giving full control to the machine. For the latter, participants stated uneasiness about relaxing their driving focus during automation especially if they expected to take back control at any moment. Some of the participants also worried about their driving skills diminishing over time as the ADS performs the driving more frequently. Technology reliability concern also appears in the MCI survey by J.D. Power and was ranked as the top concern with 70% of the respondents [8]. The same message was seen in the study by Deloitte [28], with 65% of respondents voicing a similar concern. While the industry is making big strides in the robustness and reliability of automated vehicles, the user-centric study revealed how the HMI could potentially be a key enabler in raising the confidence level in such technology and unlocking the fear about system reliability. For example, imagine an HMI that can provide situational awareness to the driver with a seamless view of what the 360-degree ADAS sensors are detecting in the car's surroundings in real time. An example HMI used in this study is shown in Figure 13, where it highlighted what the system is seeing and the safe space maintained with respect to other road users. This view was a welcome and must-have feature by several of the participants within the Anxious and Cautious profiles.

Regarding the ability to be in control, or to not relinquish it, this study found similar results as did the SAE Demo Days survey [26], with 73% preferring control to be shared between humans and the automated car. Of course, there could be multiple underlying human factors surrounding the loss of control concerns. In the user-centric clinic, the pronounced loss of control feeling was raised primarily when the vehicle could have been operating independently for a time, such as in L3, and then the user being asked to take over. Keep in mind that, at this automated level, the driver is no longer required to keep their hands on the wheel and their eyes on the road during automation. We observed that the driving transition, in particular, from automated to manual was critical at building trust in ADS. Strong HMI multimodal sensory feedback, using auditory, visual, and haptic methods - as well as monitoring the driver's state and providing a countdown for the switch back to manual - were very beneficial to prepare the driver to take back control. The takeaway about the HMI is that an effective representation of the dynamic environment and timely alerts as the driver is about to take control would be a trust builder based on responses from all four driver profiles.

The Benefit Perception The user-centric clinic also uncovered two beneficial areas, which are safety and time savings, to perform tasks that couldn't otherwise be done in a manually driven car. In comparison with the J.D. Power MCI Survey℠ [8], these two benefits were seen as third and fourth in importance, with 35% for doing other tasks and 32% for fewer crashes, respectively. User-centric participants familiar with ADAS, especially those who have it in their cars, were clear about the safety benefit. They were also imaginative when considering things to do during automated driving, such as interacting with other passengers, attending to personal needs, or being digitally immersed in their smart device. To some, the idea of unwinding and

Reprinted with permission from Valeo

FIGURE 13. Example of situation awareness view from the Valeo Mobius® cockpit.

relaxing or watching a movie was very welcoming. Similar survey findings were highlighted by the Deloitte report [28] related to interacting with other passengers, where this was rated the highest at 50%.

The User-Centric Trust Pyramid

As highlighted in Table 4, the Anxious driver was the furthest from being behind the "wheel" of an automated car, whereas the Confident driver was more eager and able to project themselves in the automated car. The HMI is a "window" to the driver in relaying what the automated car is seeing on the road. Based on the highlighted benefits and drawbacks, we can imagine a pyramid that correlates the potential trust in automation to the user experience benefits, and how the latter could be presented to the driver based on a given profile. As shown in Figure 14, the pyramid is based on three layers, and each corresponds to a certain profile mindset. At the bottom is the status quo for the current type of driving, in the middle is the "wait-and-see" phase, and at the top of the pyramid is readiness and openness about automated driving. The four profiles are then aligned according to the pyramid definition. Where Anxious and Cautious drivers do not currently trust this technology, Sporty drivers are interested, and Confident drivers showcase openness and trust. To move from one layer to the next, profile drivers need to be accompanied, reassured, and informed. The car interaction is the enabler. Shown on the right side of the pyramid are expectations inside the cockpit and behind the wheel. For the profiles with no trust in automation, we need to build their confidence based on providing them an intuitive ADAS HMI. Once they become proficient and at ease with the use of ADAS, then they can start to warm to settings customization, as in the case of the Sporty driver. Finally, once their trust in ADAS is solid, and they are capable of imagining themselves behind the wheel of an automated driving car, they would then be more relaxed and open to other non-driving activities, such as entertainment features.

Summary

Successful deployment of ADS and ultimately driverless operation vehicles depend heavily on consumer trust of the technology. Trust and other human factors are complex psychological subjects that experts in this field around the world are studying as it relates to this technology. While we are barely scratching the surface on this topic, this SAE EDGE™ Research Report analyzed and summarized results from several surveys, clinics, and studies about the consumer and, in many cases, driver acceptance of ADS.

First, the common messages on ADS are the perception of safety, as illustrated in Figure 15 from the J.D. Power 2019 MCI report [8]. Consumers are worried about the reliability of the technology, vehicle hacking, liability in case of an accident, and safety of other road users. While these are valid concerns, the technology is maturing at an impressive pace, and what we see from consumer clinics, such as the SAE Demo Days event [26], enthusiasm and trust in the technology usually increase after riding in an automated vehicle. This is consumer perception as a passenger with a vehicle operator behind the "wheel" in an ADS. From the driver's perspective, and continuing with the theme of safety, we see three groups of SAE automation levels: A) L0-L2, B) L3, and C) L4-L5. For each of these groups, the expectation on the driver versus the system might vary, and intelligent feedback monitoring of the driver to guarantee continued engagement in the driving and

FIGURE 14. The user-centric trust pyramid.

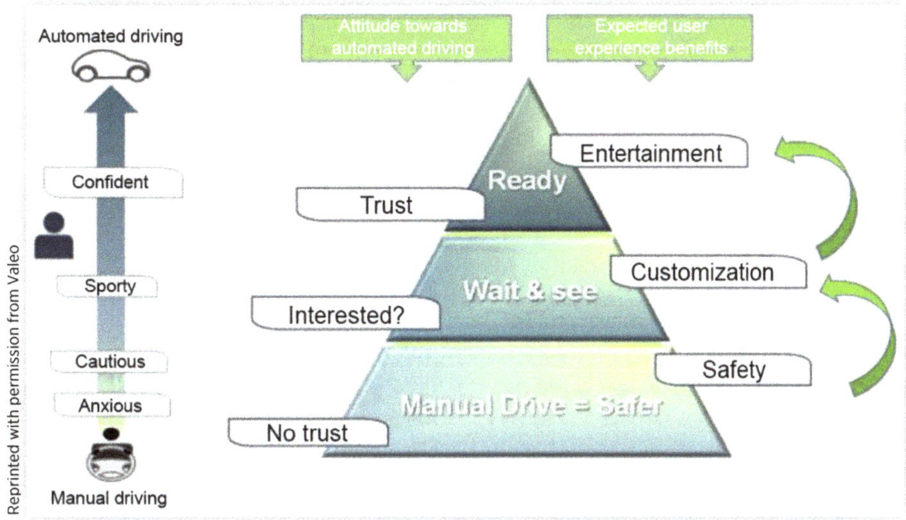

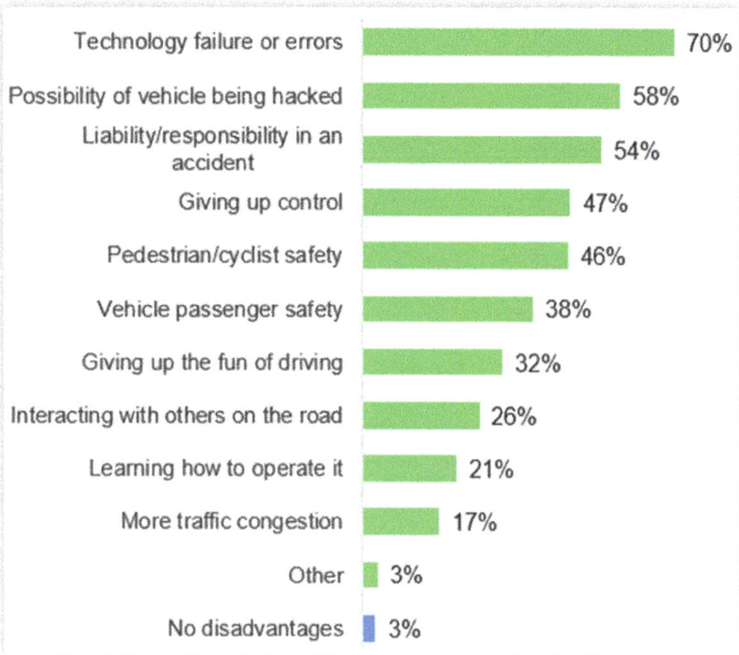

Reprinted with permission. © J. D. Power

FIGURE 15. Voice of the consumer on ADS - J.D. Power 2019 Mobility Confidence Index (MCI) Survey[SM].

being in control is addressed. Naturalistic driving studies for up to L2 are shedding light on the distraction risks and reaction time for a takeover. Lab simulator studies are finding the take-back control from automated to manual driving of about 15 seconds in the case of an L3 ADS [18, 20].

> *"The Auto-ISAC is positioned to build a global strategy in an effort to thwart the global cybersecurity threat that may impede the progress of automated vehicle development."*
>
> **Faye Francy**
> ***Executive Director, Automotive ISAC***

Second, we believe that ADAS is the building block for higher-level automation from the driver acceptance point of view. Learning from ADAS use and its optimization will be greatly beneficial to ADS. For example, how the ADAS HMI is presented to the driver needs to be consistent, logical, and intuitive in order to not annoy the driver to the point of deactivating the feature, and potentially losing its safety benefit. Understanding the limitation of the technology is also important. One idea is a standardized training process for the consumer on each of these assisted and automated driving technologies, their functionalities, and limitations. Today's drivers are seeking information on the ADAS technology from various sources, such as dealership training, owner's manual, the internet, and even trial and error [31]. Regarding training, AAA proposed a hybrid model, building upon a model from the aviation industry for partial automation [16]. The model proposes three phases, which are listed below along with the current author's summary of the primary risks in each phase.

- **Novelty phase:** Since ADAS and driving automation functions are relatively new to drivers, and they may or may not have had training on system operation, some of them might be learning about the systems through real-time use. This could increase the likelihood of safety-critical circumstances. Alternatively, drivers may not trust the system and, as a result, will not use it.
- **Post-novelty phase:** This is someone who is fully trained, but a new user. The concern is that some might have developed a preconceived notion of the system operation that is incompatible with the capabilities and limitations. In the absence of safety-critical circumstances, there is a risk of overtrust in the system.
- **Experienced-user phase:** This is a driver who has used the system repeatedly and understands it well. In such a phase, a potential work underload could manifest itself when using the system. Effects of underload, such as a driver engaging in risky non-driving tasks, including in partially automated driving, are of more concern.

Third, focusing on HMI's role as a user trust builder in ADS driving experience, four driver profiles were identified with various expectations. There were two groups for whom safety is paramount. For them, focusing on relevant information about safety would be required. For the group that still enjoys driving, ADAS or unique automated driving features would be highly desirable. Given their interest in the latest technology, they would be open to various customizable settings and forms of entertainment in their cockpit. Finally, the group of Confident drivers demonstrated a high level of experience and would be very open and trusting to be in an automated driving vehicle and, thus, would appreciate the benefits of performing other tasks than driving such as access to various forms of media entertainment. Applying the user-centric trust pyramid, we see the progression from the foundation of safety above all to the discovery stage of automated driving to being comfortable behind the "wheel" of an automated vehicle. As their trust level increases in the driving automation functions, and from low to high automation, their driving workload decreases and, as a result, their demand would increase for other non-driving tasks. Finally, the trust pyramid can further be supported by hybrid training phases introduced in the study by AAA [16].

In conclusion, the goal of this SAE EDGE™ Research Report was to provide insights on learnings gathered from various studies with the aim to gain more understanding of user acceptance of autonomy. We examined perceptions of driving assistance technologies, partial automation already on the road, ADS-fitted test vehicles, and user experiences in advanced ADS-capable cockpits. We also highlighted a means for building up users' trust in automated driving and ultimately the driverless car. It is our hope that this report is a building block for further studies as the ADS technology progresses further.

SAE EDGE™ Research Reports

SAE EDGE™ Research Reports, like the present report on "Unsettled Topics Concerning User Experience and Acceptance of Automated Vehicles," are intended to push further out into still unsettled areas of technology of interest to the mobility industry. SAE launches these reports before attempting to form a joint working group, let alone a cooperative research program or a standards committee.

SAE EDGE™ reports are intended to be quick, concise overviews of major unsettled areas where vital new technologies are emerging. An unsettled area is characterized more by confusion and controversy than established order. Early practitioners must confront an absence of agreement. Their challenge is often not to seize the high ground but to find common ground. These scouting reports from the frontiers of investigation are intended merely to begin the process of sorting through critical issues, contributing to a better understanding of key problems, and providing helpful suggestions about possible next steps and avenues of investigation.

SAE EDGE™ Research Reports, therefore, are fundamentally distinct from the more formal working groups approach and far removed from the more mature research program and standard's development process.

Next Steps for User Experience and Acceptance of Automated Vehicles

This publication should be considered only as a first step toward clarifying the issues around user experience and acceptance of automated vehicles. The intention behind this and other SAE EDGE™ Research Reports is to start a dialogue among interested parties on important industry-wide topics that require further attention. The expectation is that these explorations of unsettled areas of technology will lead to the formation of working groups and, ultimately, committees that can address and resolve the issues they raise, producing a framework for developing a common vocabulary of definitions, best practices, protocols, and standards needed to support continued progress toward safer and more innovative automotive products.

The experts' collaboration that gave rise to this publication demonstrated a great willingness on the part of the industry to define the terminology, procedures, and, eventually, the standards needed to enable an understanding of the user experience and acceptance of automated vehicles to move ahead as quickly and efficiently as possible.

This SAE EDGE™ Research Report on user experience and acceptance of automated vehicles identifies the following key topics for further pursuit, both through continued informal discussions among industry practitioners and through more formal working groups:

- Define key driver attentiveness (i.e., visual, manual, and cognitive) characteristics that require active monitoring in automated driving operations, up to L3, when the human driver is responsible for the OEDR or in the case of L3 when in DDT Fallback condition.
- Optimize the take-back control timing and standardized warning requirements for all automated to manual handover scenarios in L3 ADS.
- Expand the standardized naming concept to higher SAE automation levels for a given ODD in an automated vehicle.
- Establish a methodology for the HMI feedback requirements in automated driving vehicles, taking into account the user's experience level at a given SAE automation level.

Recommendations

The overall recommendations of this SAE EDGE™ Research Report can be summarized as follows:

1. Driver engagement sensing is required in vehicles with SAE automation levels, L0 through L2, that actively detects and ensures "eyes-on-the-road" and "hands-on-the-wheel."

2. Driver readiness sensing is required in L3-equipped vehicles to guarantee the driver is capable and ready to take back control when requested. Active feedback sensing should be capable of monitoring driver's activity/attention right before the handover, engagement during the handover, and attention after the handover and back in manual driving.
3. Occupant sensing in L4- and L5-equipped vehicles is desirable, for example, to monitor the well-being and safety of the passenger but can also be utilized for other types of monitoring (e.g., occupant count, seating position), depending on the vehicle's intended use.
4. Develop a training model, along the line of the hybrid model, for all SAE automation levels that is adaptable to the user's experience level from novice to experienced and to guarantee that the driver/user more readily embraces the technology.
5. Standardized nomenclature not only for driver assistance systems but also to higher driving automation functions.
6. Standardized feedback HMI for various SAE automation levels with minimal cognitive workload, intuitiveness, and the ability to relay situation awareness to the driver/user.
7. Support the user-centric trust pyramid model with the training model to define detailed steps in building up driver/user confidence in higher-level automation.
8. Apply the user-centric trust pyramid concept to tailor the increase of non-driving tasks during the automated driving as we gradually build up the trust in safer higher automated driving levels.

Abbreviations/Definitions

ACC - Adaptive Cruise Control.

ADAS - Advanced Driving Assistance System.

ADS - Automated Driving System. SAE J3016™ naming given to L3, L4, and L5.

AEB - Automated Emergency Braking.

APA - Automated Parking Assist.

Auto-ISAC - Automotive Information Sharing and Analysis Center, https://www.automotiveisac.com/

BSW - Blind Spot Warning.

DARPA - Defense Advanced Research Projects Agency, https://www.darpa.mil/

DDT - Dynamic Driving Task. Executing all of the vehicle operational and tactical functions for a safe vehicle motion control on the road.

DDT Fallback - The response by the user to either perform the DDT or achieve a minimal risk condition after the occurrence of DDT performance-relevant system failure or upon Operational Design Domain (ODD) exit.

Driverless Operation - SAE J3016™ naming given to L4 and L5 ADS.

FCW - Front Collision Warning.

HMI - Human-Machine Interface. An interface that enables drivers and other vehicle occupants and the car communicate intention and information.

HUD - Heads-Up Display. Display media projecting information on a transparent screen or the vehicle windshield in the driver's line of sight.

L0 - *No Driving Automation* as defined in SAE J3016™. No autonomy, a human driver performs all driving tasks.

L1 - *Driver Assistance* as defined in SAE J3016™. Vehicle controlled by a human driver, but some driving assist features might be included in the vehicle design.

L2 - *Partial Driving Automation* as defined in SAE J3016™. Vehicle has combined automated longitudinal and lateral motion control, but a human driver must remain engaged with the driving task at all times.

L3 - *Conditional Driving Automation* as defined in SAE J3016™. The vehicle is capable of performing the driving functions under certain conditions, but a human driver must be present and attentive at all times for a fallback control when requested.

L4 - *High Driving Automation* as defined in SAE J3016™. The vehicle is capable of performing all driving functions under certain conditions. A human driver has no fallback role but may have the option to take control if desired.

L5 - *Full Driving Automation* as defined in SAE J3016™. The vehicle is capable of performing all driving functions under all conditions. A human driver has no fallback role but may have the option to take control if desired.

L2NDS - L2 Naturalistic Driving Studies. Naturalistic driving studies investigate real traffic situations and provide useful information for critical events, near crashes, or actual ones, as well as normal driving in mixed traffic conditions.

LDW - Lane Departure Warning.

LiDAR - Light Detection And Ranging. An active sensor that measures the distance to a target by illuminating the target with light, typically a laser source, and measuring the reflected light onto the sensor receiver.

MRM - Minimal Risk Maneuver. (Synonymous with Minimum Risk Condition) Driving maneuver that the ADS-equipped vehicle must execute to take it from the current driving condition to a safe stop. In L3, the maneuver could happen due to the human driver's unresponsiveness to the takeover request or system failure.

NHTSA - National Highway Traffic Safety Administration, https://www.nhtsa.gov/

NSC - National Safety Council, https://www.nsc.org/

ODD - Operational Design Domain. A specified scenario in which an automated vehicle can safely operate on its own.

OEDR - Object and Event Detection and Response. Subtasks of the DDT that include monitoring the driving environment and executing an appropriate response to the detected objects and events.

RCTW - Rear Cross Traffic Warning.

RTI - Request To Intervene.

RADAR - RAdio Detection And Ranging. Similar to a LiDAR in using radio wave source.

Ultrasonic - An active sound wave sensor, using ultrasonic frequency, for measuring the distance to targets based on the evaluation of the returned echoes to the sensor.

U.S. DOT - United States Department of Transportation, https://www.transportation.gov/

Acknowledgments

The author wishes to thank all of the contributors listed in this SAE EDGE™ Research Report, for the valuable information and exchanges leading to the creation of this report. Special thanks also go to Mary Moore and Marcie Hineman at SAE International for their valuable insights on the SAE Demo Days event.

Greg Brannon, *AAA*
Nicolas De Crémiers, *Navya*
Alex Epstein, *National Safety Council*
Rachel Forestier, *Valeo*
David Hofert, *Perrone Robotics*
Katharina Hottelart, *Valeo*
Kristin Kolodge, *J.D. Power*
Sheldon Russell, Ph.D., *Virginia Tech Transportation Institute*

The authors of this document together with the SAE Team responsible for its creation join in expressing our deepest appreciation to all the individuals mentioned above.

Amine Taleb-Bendiab, Ph.D.
SAE International masthead

References

1. Dormehl, L. and Edelstein, S., “Sit Back, Relax, and Enjoy a Ride through the History of Self-Driving Cars,” Digital Trends, February 2019, https://www.digitaltrends.com/cars/history-of-self-driving-cars-milestones/
2. Wikipedia, “History of Self-driving Cars,” Wikipedia, Latest Update: January 18, 2020, https://en.wikipedia.org/wiki/Self-driving_car
3. Defense Advanced Research Projects Agency (DARPA), “The DARPA Grand Challenge - Ten Years Later,” March 2014, https://www.darpa.mil/news-events/2014-03-13
4. Baker, S., “Unsettled Topics Concerning Sensors for Automated Road Vehicles,” SAE EDGE™ Research Report, October 2018.
5. SAE J3088TM, *Active Safety System Sensors* (Philadelphia, PA: SAE International, November 2017).
6. Abuelsamid, S., “NXP And Kalray Team Up On High-Performance Automated Driving Compute Platform,” Forbes, January 2019, https://www.forbes.com/sites/samabuelsamid/2019/01/08/nxp-and-kalray-team-up-on-high-performance-automated-driving-compute-platform/#3178b78c6ee9
7. U.S. Census Bureau, 2014-2018 American Community Survey 5-year Estimate - “Mean Travel Time to Work - USA,” U.S. Census Bureau - 2018, https://www.census.gov/search-results.html?q=commute+time+to+work+2019&page=1&stateGeo=none&searchtype=web&cssp=SERP&_charset_=UTF-8
8. J.D. Power Mobility Confidence Index - “Most Consumers Still Say ‘Thanks, but No Thanks’ to Future Self-Driving and Electric Vehicle Offerings,” J.D. Power, Press Release, January 2020, https://www.jdpower.com/business/press-releases/2019-q4-mobility-confidence-index-study
9. SAE J3016™, “Taxonomy and Definitions for Terms Related to Driving Automation Systems for On-Road Motor Vehicles,” SAE International, June 2018 (Revised).
10. Reimer, B. et al., “An Evaluation of Driver Reactions to New Vehicle Parking Assist Technologies Developed to Reduce Driver Stress,” MIT UTC / MIT AgeLab, November 2010.
11. Reimer, B. et al., “Reductions in Self-Reported Stress and Anticipatory Heart Rate with the Use of a Semi-Automated Parallel Parking System,” *Applied Ergonomics* 52:120-127, January 2016, https://doi.org/10.1016/j.apergo.2015.07.008.
12. DOT HS 812 517 - Traffic Safety Facts - “Distracted Driving,” NHTSA, April 2018, https://crashstats.nhtsa.dot.gov/Api/Public/ViewPublication/812517
13. Deeken, A., “Safe to Use Hands-Free Devices in the Car? Yes, According to Research,” Virginia Tech Daily, February 7, 2019, https://vtnews.vt.edu/articles/2019/02/020519-vtti-handsfreestudy.html
14. Dingus, T.L. et al., “The Prevalence of and Crash Risk Associated with Primarily Cognitive Secondary Tasks,” *Safety Science* 119:98-105, November 2019.
15. DOT HS 812 642, “Naturalistic Study of Level 2 Driving Automation Functions,” NHTSA, November 2018.
16. Dunn, N. et al., VTTI - “Understanding the Impact of Technology: Do Advanced Driver Assistance and Semi-Automated Vehicle Systems Lead to Improper Driving Behavior?,” AAA Foundation for Traffic Safety, December 2019, https://aaafoundation.org/wp-content/

uploads/2019/12/19-0460_AAAFTS_VTTI-ADAS-Driver-Behavior-Report_Final-Report.pdf

17. Notice of Federal Guidelines, "Visual-Manual NHTSA Driver Distraction Guidelines for In-Vehicle Electronic Devices," NHTSA-2010-0053, Federal Register Vol. 78, No. 81, April 2013.
18. Thatcham Research and ABI, "Defining Safe Automated Driving," September 2019.
19. Merat, N. et al., "Transition to Manual: Driver Behaviour When Resuming Control from a Highly Automated Vehicle," *Transportation Research Part F* 27:274-282, October 2014.
20. Kuehn, M. et al., "Takeover Times in Highly Automated Driving (Level 3)," in *25th International Conference on the Enhanced Safety of Vehicles*, Detroit-Michigan, 17-0027, June 2017.
21. DOT HS 812 182, "Human Factors Evaluation of Level 2 and Level 3 Automated Driving Concepts," NHTSA, August 2015.
22. Valeo Mobius®, https://www.youtube.com/watch?v=yxx81s1YE4E
23. Diederichs, F. et al., "New HMI Concept for an Intuitive Automated Driving Experience and Enhanced Transitions," in *AutomotiveUI Conference*, Nottingham, UK, ACM 978-1-4503-3736-6, September 2015.
24. Reilhac, P. et al., *User Experience with Increasing Levels of Vehicle Automation: Overview of the Challenges and Opportunities as Vehicles Progress from Partial to High Automation* (Springer Int'l Publishing AG, 2017), 457-482.
25. Mcity, University of Michigan, "Mcity Driverless Shuttle - A Case Study," Mcity, Ann Arbor-Michigan, June 2018.
26. SAE, "SAE Demo Days Survey," SAE International, Press Release, November 5, 2019, https://www.sae.org/news/press-room/2019/11/sae-international-survey-shows-public-enthusiasm-about-self-driving-cars
27. Wadsworth, J., "Lessons Learned from City of Las Vegas Driverless Shuttle Pilots," City of Las Vegas, Nevada, May 2018 (unpublished).
28. Deloitte, "Autonomous Driving in Germany - How to Convince Customers," Analytics Institute, May, 2017, https://www2.deloitte.com/content/dam/Deloitte/de/Documents/consumer-industrial-products/Autonomous-driving-in-Germany_PoV.pdf
29. Edmonds, E., "Three in Four Americans Remain Afraid of Fully Self-Driving Vehicles," AAA, NewsRoom, March 2019, https://newsroom.aaa.com/2019/03/americans-fear-self-driving-cars-survey/
30. Edmonds, E., "AAA Recommends Common Naming for ADAS Technology," AAA, NewsRoom, January 2019, https://newsroom.aaa.com/2019/01/common-naming-for-adas-technology/
31. McDonald, A. et al., The University of Iowa - "Vehicle Owners' Experiences with and Reactions to Advanced Driver Assistance Systems," AAA Foundation for Traffic Safety, September 2018, https://aaafoundation.org/wp-content/uploads/2018/09/VehicleOwnersExperiencesWithADAS_TechnicalReport.pdf
32. J.D. Power Consumer Tech Experience Index, "Customer Demand for Safety Technology Threatened by Overbearing Alerts," J.D. Power, Press Release, August 2019, https://www.jdpower.com/business/press-releases/us-tech-experience-index-txi-study
33. Laris, M., "The Federal Government Asked for Advice on Self-Driving Cars. It Got a Mouthful," The Washington Post, April 8, 2016, https://www.washingtonpost.com/news/dr-gridlock/wp/2016/04/08/the-federal-government-asked-for-advice-on-self-driving-cars-it-got-a-mouthful/
34. Hersman, D., NSC - "The Automated & Self-Driving Vehicle Revolution: What is the Role of Government?," NSC, November 16, 2016, https://www.nsc.org/Portals/0/Documents/NSCDocuments_Corporate/Speeches%20and%20Testimony/THUD-Testimony-111616.pdf
35. "Clearing the Confusion: Recommended Common Naming for Advanced Driving Assistance Technologies," AAA+CR Reports+J.D. Power+NSC Joint Report, October 2019, https://www.safetyandhealthmagazine.com/articles/19172-nsc-part-of-effort-to-standardize-names-of-advance-driver-assistance-technology
36. Transportation.gov - "U.S. Transportation Secretary Elaine L. Chao Announces New Initiatives to Improve Safety on America's Roads," U.S. DOT, Briefing Room, January 2020, https://www.transportation.gov/briefing-room/us-transportation-secretary-elaine-l-chao-announces-new-initiatives-improve-safety

Contact Information

EDGEresearch@sae.org.

The authors of this document together with the SAE Team responsible for its creation join in expressing our deepest appreciation to all of the individuals who contributed.

www.ingramcontent.com/pod-product-compliance
Lightning Source LLC
LaVergne TN
LVHW070912210926
848482LV00017B/1062
* 9 7 8 1 4 6 8 6 0 1 8 7 9 *